GEOLOGIC
Processes

Teacher Supplement

GEOLOGIC *Processes*

Teacher Supplement

Institute for Creation Research

SCIENCE EDUCATION ESSENTIALS is a curriculum supplement series designed to cover vital topics in the various science disciplines, all from a thoroughly biblical viewpoint. Each product includes a teacher instructional guide, along with K-12 activities and classroom helps to guide discussion, reinforce subject content, and facilitate hands-on laboratory exercises.

Published by the Institute for Creation Research.

SCIENCE EDUCATION ESSENTIALS

Series Creator:	Dr. Patricia L. Nason
Project Manager:	Janis McCombs
Managing Editor:	Beth Mull
Assistant Editor:	Christine Dao
Graphic Designer:	Susan Windsor
Science Reviewers:	Dr. Charles McCombs, Dr. Jeff Tomkins, Dr. Randy Guliuzza, Dr. Chris Osborne, Dr. Larry Vardiman, Dr. Tim Clarey, Dr. Brad Forlow, Brian Thomas, Frank Sherwin

ISBN: 978-0-932766-96-0

GENETIC DIVERSITY

Teacher Supplement Author:

Dr. John Morris, perhaps best known for leading expeditions to Mt. Ararat in search of Noah's Ark, served on the University of Oklahoma faculty before joining the Institute for Creation Research in 1984. He received his Doctorate in Geological Engineering at the University of Oklahoma in 1980. Dr. Morris held the position of Professor of Geology before being appointed President in 1996. He travels widely around the world speaking at churches, conferences, schools, and scientific meetings. Dr. Morris has written numerous books and articles on the scientific evidence that supports the Bible. Dr. Morris is the author or co-author of such books as *The Young Earth, The Modern Creation Trilogy,* and *The Fossil Record: Unearthing Nature's History of Life*.

K-12 Instructional Contributors: Dr. John Morris, Dr. Patricia Nason, Dr. Charles McCombs, Janis McCombs, Leona Criswell, Brian Thomas

For additional resources from the Institute for Creation Research, please visit www.icr.org or call 800.337.0375.

Copyright © 2010 by the Institute for Creation Research. All rights reserved. No portion of this book may be used in any form without written permission of the publisher, with the exception of brief excerpts in articles and reviews. For more information, write to Institute for Creation Research, P. O. Box 59029, Dallas, TX 75229.

ISBN: 978-1-935587-05-7

Printed in the United States of America.

Table of Contents

	PAGE
Preface	7
Introduction	9
The Earth as a Result of Past Processes	13
Earth's Present Makeup	15
Types of Rock in the Crust	16
Divisions within the Crust and Upper Mantle	17
Geologic Processes Acting Today	21
Erosional Processes That Shape the Earth's Surface	21
Tectonic Forces That Shape the Earth	25
Depositional Processes That Shape the Earth	27
The Deposition of Igneous Rock	27
The Deposition of Sedimentary Rock	32
Oceanic Processes That Shape the Earth	36
Creation and the Genesis Flood	39
The Great Flood of Noah's Day	40
Conclusion	43
Bibliography	45

PREFACE

Teachers mold the minds of their students, helping them construct knowledge and an understanding of the world around them. A teacher's influence on the belief system, as well as cognitions, of a student can affect the student for a lifetime.

For nearly 40 years, the Institute for Creation Research has equipped teachers with evidence of the accuracy and authority of Scripture. In keeping with this mission, ICR presents Science Education Essentials, a series of science teaching supplements that exemplifies what ICR does best—providing solid answers for the tough questions teachers face about science and origins.

This series promotes a biblical worldview by presenting conceptual knowledge and comprehension of the science that supports creation. The supplements help teachers approach the content and Bible with ease and with the authority needed to help their students build a defense for Genesis 1-11.

Each science teaching supplement includes:

- A content book written at the high school level to give teachers the background knowledge necessary to teach the concepts of scientific creationism with confidence. Each content book is written and reviewed by creation scientists, and can be purchased separately in class sets.

- A CD-ROM packed with teacher resources, including K-12 reproducible activities and PowerPoint presentations. The instructional materials have been pilot tested for ease in following instructions and completeness of activities. They have also been reviewed by scientists for scientific accuracy and by theologians for biblical correctness.

Science Education Essentials are designed to work within a school's existing curriculum, with an uncompromising foundation of creation-based science instruction. Secular textbooks are finding their way into Christian schools. Teachers may not lack belief in the Word of God, but they often do not have adequate information or knowledge concerning the tenets of scientific and/or biblical creation. Science Education Essentials equips teachers with the tools they need to teach the science of origins from a biblical rather than an evolutionary worldview.

The goal of each science supplement is to:

a) increase the teacher's understanding of and confidence in scientific creation and the truth of God's Word, while glorifying God as Creator;

   But sanctify the Lord God in your hearts: and be ready always to give an answer to every man that asketh you a reason of the hope that is in you with meekness and fear. (1 Peter 3:15)

b) provide teachers with a toolkit of activities and other instructional materials that build a foundation for their students in creation science apologetics;

Beware lest any man spoil you through philosophy and vain deceit, after the tradition of men, after the rudiments of the world, and not after Christ. (Colossians 2:8)

c) encourage the use of the higher level thinking necessary to stand firm against the lies of evolution and humanism.

…that we henceforth be no more children, tossed to and fro, and carried about with every wind of doctrine, by the sleight [trickery] of men, and cunning craftiness, whereby they lie in wait to deceive; but speaking the truth in love, may grow up into him in all things, which is the head, even Christ. (Ephesians 4:14-15)

With Science Education Essentials, teachers can equip the future generation of scientists and individuals to examine the evidence for the truth of Scriptures through an understanding of creation science. By using hands-on activities and relating scientific truth to the Bible, teachers/parents will be grounding their children in creation science truths so they can provide a logical response when challenged with science that is based on a philosophy that is in direct contradiction to Genesis 1-11.

As the leading creation science research organization, ICR is providing meaningful creation science material for classroom use. Our desire is that the materials renew minds, defend truth, and transform culture (Romans 12:1-2) for the glory of the Creator.

Dr. Patricia L. Nason

Introduction

*For thus saith the L*ORD *that created the heavens; God himself that formed the earth and made it; he hath established it, he created it not in vain, he formed it to be inhabited: I am the L*ORD*; and there is none else. (Isaiah 45:18)*

The scientific community generally holds that the earth was formed billions of years ago and that the geologic formations we see today are the result of millions of years of natural geologic processes. The Bible, however, directly contradicts this assessment, instead presenting the earth as having been created and populated by God in six 24-hour days just thousands of years ago. Does the earth itself provide any evidence that can answer the question of which of these views is correct?

Actually, yes—quite a bit of evidence, as can be illustrated by an experience I had several years ago in Beijing, China. The incident dated from the days just prior to China's cracking open its doors to outside influence. It started with the reunion of a brother and sister who had not seen each other since World War II. Their meeting involved both high-ranking members of China's central government and an American pastor, and climaxed with a promise for a seminar in China on the Christian worldview. The question to be addressed: How would China benefit if Christianity were allowed to enter?

The Institute for Creation Research, known for its uncompromising stance on creation and Christian thinking, was asked to take the lead in organizing the seminar and recruiting the speakers. The Chinese government promised an audience that was commensurate with the prestige of the speakers. Fully realizing the potential, I dove into the task of identifying prominent Christian politicians, educators, economists, scientists, medical doctors, and others committed to Christian thinking in their fields. I prepared guidelines and short abstracts to ensure that all important topics were covered and that there was

Temple of Heaven, Beijing, China

no subject duplication. We were required to submit advance copies in writing of our proposed lectures, along with our resumes and travel documents.

That is when the difficulties started. One by one, the speakers were denied visas into the country. Soon we were left with just two of the originally approved speakers: a Christian astronaut who had been born in China to Christian missionaries—and me. This was probably orchestrated to "save face," since I had been the primary planner of the now-aborted seminar. But the night before the seminar, I was roused from sleep and informed that I would not be speaking, either.

Intense negotiations began with several top Chinese officials over my talk, of which they had the submitted transcript. They absolutely forbade any mention of Christianity, Christ, creation, the Gospel, or any spiritual content. It was turning out to be less and less of a conference on the Christian worldview. Finally, I promised to mention only my own personal geologic research, and they agreed.

There was no time to clean up or get breakfast. I quickly changed clothes, set up the slide projector, and briefed my interpreter. The auditorium was filled with 2,500 scientists, graduate students, politicians, and government workers. I stuck to the agreement. I spoke only of my personal geologic research—on the recent eruption of Mount St. Helens.

Mount St. Helens

I explained that when Mount St. Helens erupted, the melting glacier at the summit sent cascades of water rushing down the mountain, gathering mud, rocks, and trees, eventually coming to rest in great blankets of sediments. These sediments hardened over the next few years into sedimentary rock, complete with various fossils. Volcanic ashfall buried the pristine forest. Coalification and petrification of the trees commenced immediately. Subsequent erosion formed mature landscapes, complete with drainage patterns and canyons. Radioisotope dating of the resulting igneous products also pointed to great age. All factors indicated that the eruption must have taken place long, long ago, yet we had witnessed the volcanic activity take place only recently.

I explained the significance of our analyses and how they were contrary to the great ages for the earth taught to students. Like them, my education had censored the truth of rapid geologic processes and the fallacy of geologic dating schemes. Following my formal education, I had only been able to repeat past errors. Only as I began to research these topics myself did I develop useful skills in geology. "Censorship never does any good," I repeated to my audience. "Censorship of the truth made me an inferior geologist."

Surely, the earth's surface has been sculpted by geologic processes, but at what rate, scale, and intensity? We must follow the laws of nature, like gravity and radioisotope decay, but are we justified in limiting

Mount St. Helens mudflow

GEOLOGIC PROCESSES

them to today's observations? Normally, geologists blindly assume that natural processes acted in the past much as they do today. All other information is "censored" from students in the name of uniformitarian naturalism. Incorporating knowledge of catastrophic processes into our thinking, however, often allows a better geologic interpretation of events that happened in the unseen past.

This curriculum supplement will discuss processes that shape the earth today, as well as the processes that shaped the earth in the unseen past. As you will see, if we restrict the processes that operated in the past to only the rates that are observed today, we become incapable of explaining the geologic record. Greater, more rapid and dynamic processes are necessary. By including information that is usually censored, students will be better prepared to understand geologic data.

The Earth as a Result of Past Processes

The planet we live on is a product of past processes. As far as we know, no planets like ours have formed anywhere else in the solar system or in nearby star systems. Irregularities in the rotation of distant stars imply that they are circled by planets, but these bodies are not like earth. Typically they are gas giants, more like Jupiter, that are totally incapable of supporting life.

Earth seems to be quite unique and supports an abundance of life. What past processes formed it? Could present processes have accomplished the job? Or are other processes—or the same processes operating at rates, scales, and intensities not seen today—required?

The majority of scientists think the earth started as a collection of star dust spewed out by past exploding stars. Most of the material was lightweight hydrogen and helium that was collected by gravity to form the sun, our star. The leftover, heavier atoms collected into planets, with the earth-like planets near the sun and the gaseous planets farther out. Internal processes and the immense gravity of the collecting sphere caused the earth to heat to the melting point. This molten ball cooled and gained water over time, retaining the excessive heat only in its interior. Rocks formed as the molten material cooled and solidified, dividing into layers. This hypothetical melting of the early earth is known as the Iron Catastrophe. But was our planet actually formed this way? Let's examine the earth to see what it tells us.

View of earth from the moon

GEOLOGIC PROCESSES

Earth's Present Makeup

In order to investigate whether present processes could have been responsible for earth's formation, we must first understand its present form. Planetary scientists divide it into three main sections, a series of concentric spheres, one within the other.

The thin veneer on the surface is called the *crust*, on which we live. It is composed of silicate minerals, like quartz and feldspars. All the mountains, continents, and ocean basins are part of the crust. Even though its thickness only measures a few miles, scientists have never broken through to the layers beneath. Neither the deepest oil well nor the deepest oceanic trench penetrates the crust. Volcanic eruptions usually only recycle crustal material that has been forced downward, heated, and spewed back out. We are still learning about the earth's internal character, particularly regarding the various zones within it.

The thick *mantle* underlies the crust and comprises about 80 percent of the planet. It seems to be made of slightly different materials, richer in iron and magnesium, than the crust, but no one has been there to make an observation. Partial melts of the upper mantle produce basalt-rich rocks like those that are found at the mid-ocean ridges. Some enigmatic eruptions of the past were quite different from modern volcanic eruptions and have extremely deep sources for the erupted material, such as in the diamond-bearing diatremes and kimberlite pipes.

At the earth's center we find the *core*, which is further divided into two parts: the *inner core* and the *outer core*. Both parts of the core are thought to be made of compressed metal, probably iron and nickel. By measuring the way seismic waves pass through the earth, we infer that the outer core is a liquid and the inner core is a solid. The outer core remains a metallic liquid because it is under less pressure than the inner core. The inner core is probably even hotter than the outer, but because of its greater pressure, it has been squeezed into a solid.

Keep in mind that scientists do not really know how a solid reacts to sustained high temperatures and pressures like this, and we have no direct knowledge of the core's makeup, density, or temperature. We can only surmise its true nature from indirect measurements. For instance, we do know the total mass and volume of earth, but must estimate what its component parts contribute to the whole. This can be done with a measure of confidence, but some uncertainty still exists.

All of man's raw materials for industry and culture come from the crust, and the processes that operate there will dominate the discussions of this book. Those influences from without (such as the oceans' tides, caused primarily by the moon's gravity) or deep within (such as the magnetic field surrounding the earth, evidently due to electric currents in the core) are not insignificant, but will be left to other treatments.

Types of Rock in the Crust

Three general types of rock can be found on earth's surface, each with numerous subcategories.

Igneous. Great heat exists within the deeper layers of the planet—great enough even to melt rock and turn it into molten magma. In liquid form, the magma can erupt onto the surface in volcanic events as lava, be squeezed into cracks in overlying solid rocks as dikes, or be injected as a large "bubble" within rocks as a pluton, sometimes comprising the core of mountains. Igneous rocks form as the hot, liquid magma "freezes," or solidifies, allowing crystals to grow from the melt. Extremely dense igneous rocks that formed in the ocean depths as hot lava flows are known as *basalts*, while less dense igneous rocks (*granites*) comprise the bulk of the continents. These rocks are eroded to eventually become sedimentary deposits and sedimentary rock.

Basalt field on Ararat in Turkey

Sedimentary. Sediments are deposited on the earth's crust by various processes, and can then harden into sedimentary rock. The most common depositional environment for sediments is in the ocean, where sediments collect and can later be uplifted to become continental rock. Sedimentary rocks form a thin blanket over the continents. One type of sedimentary rock is *clastic* rock. These consist of redeposited particles eroded from previously existing rock. Sandstone is a clastic rock consisting of sand-sized particles that originated in granite. Shale is clastic rock consisting of even smaller, clay-sized particles. Sedimentary rocks may also be chemically deposited, their solids having been dissolved by water and later precipitated out of the water when conditions changed. Limestone formed this way, and so do stalactites.

Crossbedding in Coconino Sandstone (note person for scale)

Metamorphic. Sometimes rocks are subjected to extreme levels of heat and pressure, and are altered into a completely different kind of rock without having melted. Just as a caterpillar can metamorphose into a butterfly and appear completely different, so too can limestone metamorphose into marble, shale into slate, etc. The original chemicals remain the same, but they morph (recrystallize) into another form, often producing new minerals and textures. Both sedimentary and igneous rocks will metamorphose when conditions are extreme.

Divisions within the Crust and Upper Mantle

Previous generations of geologists thought that the continents were stable and that they occupied the same positions they always had. Now the majority agree that the continents are divided into "plates" that move with respect to one another. Plates consist of crust, whether continental or oceanic, and the uppermost mantle to a depth of 60 to 100 miles. Some plates move apart, leaving a void that is filled in by new basaltic, oceanic crust. The Atlantic Ocean is thought to have formed

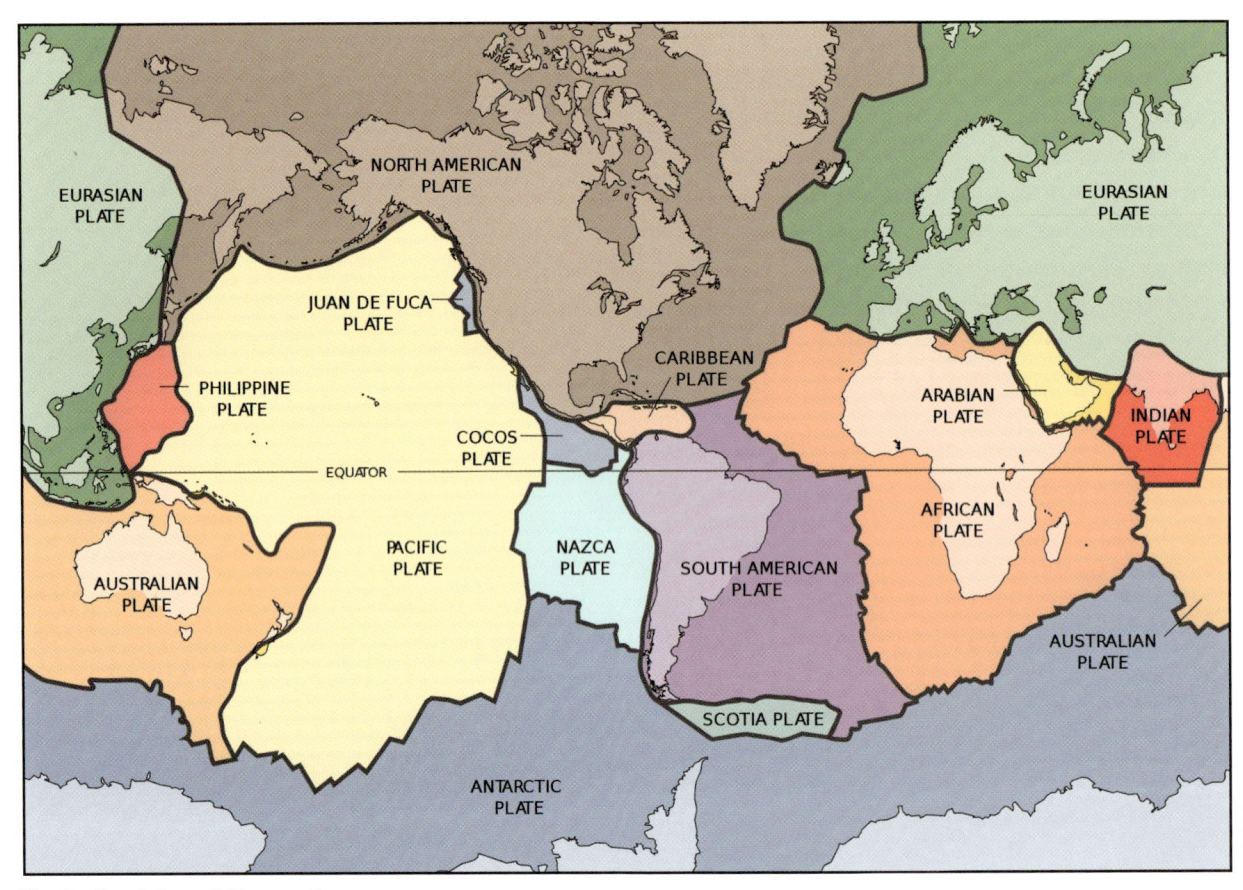

Tectonic plates of the earth

in this way as the Americas moved away from Europe and Africa. Other plates slip laterally beside each other, such as along the famed San Andreas Fault in California. One side is moving northward with respect to the other. When continental plates move toward each other, such as when India collided with Asia, large mountains crumple up. When a continental plate collides with an oceanic plate, the more dense basaltic oceanic plate sinks beneath the lighter continental plate, which in essence "floats" over the oceanic plate. The ocean material sinking below the continent is heated, melts, and rises to the surface again through periodic volcanic eruptions.

Today, the continents are quite stable, hardly moving at all, but it is believed that they were all once conjoined. Evidence to support the idea that they were once connected comes from the rather obvious coastline match-up of some of the continents, such as Africa and South America. More importantly, the overlying sediment layers also testify to this, for the same series of strata can sometimes be found on either side of a wide ocean, indicating that the continents were connected at the time those sediments collected, but were later separated.

The current plate boundaries are identified by plotting the worldwide locations of earthquakes. Differential movements of the plates, whether diverging (moving apart), converging (moving together), or moving laterally, would be accompanied by earthquakes focused in discrete linear areas. These active zones define the plate boundaries and typically point to regions of greater volcanic activity. For instance, around the huge Pacific Plate, which is essentially the Pacific Ocean, are active volcanoes that have collectively been dubbed the "Ring of Fire." Deformed and folded mountain chains are often associated with these volcanoes, testifying to past and present plate activity.

If the continents could be moved back together, which can be simulated on a computer, they would seem to fit with only a few minor overlaps or gaps. This has led some to propose that a single super-continent called Pangaea became divided into our present geography. Exactly when this happened and the details of the past plate movements are still being researched, a process that is made difficult by the obvious fact that the earlier evidence has been smeared and erased by later plate activity. Discerning the past from the partial and scanty evidence that remains in the present is a difficult job, but scientists do their best.

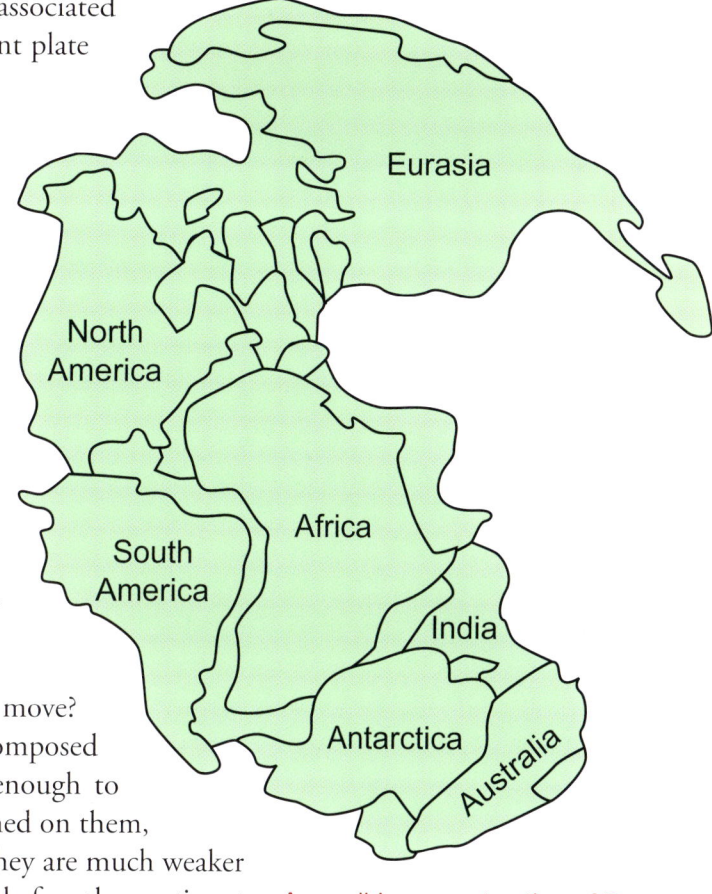

Many have wondered—just how do continents move? Plates are immense. The rocks of which they are composed may seem hard to us, but they are not strong enough to withstand movement across the earth. If you pushed on them, they would crush before the continent budged. They are much weaker if you pull on them—they would break apart long before the continent would move. This is a big problem!

A possible reconstruction of Pangaea

This has led some to suggest that a special layer of hot, plastic-like material exists at the base of the plate that allows movement of the overlying, more rigid plate. Obviously, this layer has not been directly observed, but many think it is necessary to move the continents. It has been indirectly identified by studying the movement of earthquake waves through the upper mantle. All seismic waves slow down as they pass through this "mushy" solid, called the asthenosphere. We must keep in mind that while plate movements seemingly answer a lot of questions about the past, we still don't understand how they move using present processes operating at present rates, scales, and intensities.

GEOLOGIC PROCESSES

Geologic Processes Acting Today

Now that we know something of the earth's present makeup, we can discuss present processes that work to alter it. Processes of the past are the subjects of history and scientific theory, and even science fiction, but modern processes are observable and measurable. Experiments can be used to test and employ them. These are the tools of science and the domain of the empirical scientist.

Erosional Processes That Shape the Earth's Surface

Most easily observed are processes of erosion. Many people have discovered that leaving a garden hose on overnight will create a rut through their yard. A large rainstorm can send sheets of water over a soil surface or a parking lot, doing major damage. Arroyos are stream beds that are usually dry, but can flash-flood from rain that falls upstream, even at some distance away. Unless anchored down, plants, animals, people, and even buildings can be swept away by the power of cascading water during a flood.

We have all been told that Grand Canyon was gradually eroded by the Colorado River over long, long ages. Today, the canyon measures up to eighteen miles wide and over a mile deep. Yet the river accomplishes very little erosion on a normal day. During a wet period, side canyon streams carry a lot of silt, and occasionally boulders, down to the river,

Grand Canyon. The river is much too small to have eroded such a large canyon.

but the river channel itself experiences very little erosion. Careful studies have shown that the river bottom is normally covered by a layer of gooey mud that prevents the moving water from even encountering the underlying rock unless the water volume and velocity reach a critical level that is sufficient to remove the mud and begin to pluck off pieces of the bottom rock. At flood stage, great geologic work can be accomplished, and was accomplished before dam construction controlled the river flow. Calculating an average erosion rate compounds the error, for it is based on an assumption about past storm frequency and ferocity. What if the past was different, including climate and rainfall levels?

A hint of Grand Canyon erosion rates can be seen by noticing that virtually all of the canyon's walls are covered by a thin layer that is different from the rock itself. For instance, the Redwall Limestone, a prominent horizontal layer about midway between the bottom and the top, sports a beautiful red color. But a fresh surface of the rock is gray, not red. An exposed surface is stained red by water washing over the gray layer from the rusty Hermit Shale above and by the wind blowing particles around. Measurements today show that at the rate the coloring is acquired, it would take a few thousand years to reach its present color. Obviously, the rock has not been significantly eroded (exposing new surfaces) for thousands of years or else it would not have retained its distinctive red color. It seems that almost no significant erosion is going on today.

The Colorado River's red or brown color comes from eroded material, yet the majority of this is not from the canyon itself but instead washes downstream from the mountains of Colorado and Utah. One of Grand Canyon's big mysteries is the location of the vast amount of sediment that would have been removed from the Colorado Plateau if it were extremely old. Present river velocities, even during river flooding, would deposit sediment locally when the river slowed down. However, no nearby large deposit has been found as predicted. There is a similar sediment deposit far away, in southern California, but it would take sustained flooding on a scale not currently observed to produce it.

When viewed from the air, the Colorado River can be seen as a single gouge in an otherwise flat, featureless plain. Remnants of the overlying, but now missing, strata can still be seen in a few places, so we know they once existed. But now there is only a plateau. Standard thinking considers the plateau to have been in existence for long ages before the river began to erode through it. If so, where are the side canyons and the mature drainage patterns that would have been established along such a river? Today, the only way this sheet-like appearance can be accomplished in a drainage area is with deep water flowing everywhere at high velocities. The modern river valley dates from after the sheet erosion episode. It looks like the river valley is the result of processes that did not operate at the same slow paces they do today.

Aerial view of Grand Canyon. Rapidly-flowing water covering this entire region was needed to form the flat plateau. As the water flow lessened, it channelized and formed the large canyon. Now only a "trickle" remains.

A well-known worldwide observation is that of "under fit rivers." Nearly every river valley that exists today is much larger than what the present river system requires or could erode. Charles Darwin made a classic blunder when he encountered such a river valley in southern Argentina. He assumed that the river, acting slowly over long ages, was responsible for the canyon, when actually major ice-melting episodes carrying large volumes of water at high velocities carved it out in spurts. His erroneous way of thinking in geology later yielded disastrous results when applied to biology and the origin of species.

This long-age thinking also comes into play at Mount St. Helens. The sudden, rapid heating by the exploding volcano quickly melted the glacier on the mountain's summit, sending torrents of water and mud racing down the slopes. Erosion of the underlying sediments was extensive, and even hard layers of solid lava from previous eruptions were quickly gouged out. Elsewhere, canyons that looked just like a miniature Grand Canyon were eroded in a short period of time. Under the right conditions, with rapidly moving water and excessive energy levels, extensive erosion can occur quickly.

Water has great erosional power in more than one form, because frozen water, in the form of glaciers, can efficiently scour soil and rock. The moving, hard ice traps grit, rocks, and boulders that act as sandpaper as the glacier creeps along at "glacial" speeds, sometimes leaving tell-tale grooves in the rock in the direction of flow. The moving body of ice scrapes the rocks adjacent to it and plucks up large pieces from below. A moving continental glacier scrapes the rock off everywhere, while a

GEOLOGIC PROCESSES

descending mountain glacier carves out a U-shaped valley, as opposed to the V-shaped valley normally left by river erosion. The eroded pieces of rock are incorporated into the ice and are deposited when the glacier comes to a halt or melts.

Ice has a greater volume than water, thus water swells when it freezes. This means that water within the pores of rocks expands and exerts a tremendous pressure on the rock that contains it. If this pressure exceeds the rock's strength, the rock will break and erosion will proceed more rapidly. Called freeze-thaw weathering, it can eventually break down almost every type of rock.

Similarly, seeds and plants can wedge their way into cracks in rocks. As they grow, they take up more space and can force even mighty rocks to give way. In this way, tiny roots can crack a sidewalk, eventually causing it to deteriorate. Even little animals, such as burrowing moles, insects, or earthworms, can assist in breaking up hard rock.

Wind can also accomplish significant erosion, especially in places where the wind direction is constant. Once small particles are carried aloft, they contribute to the power of the pelting wind. Some rocks have less resistance and are more susceptible to this erosive action. The windborne grit often leaves hollowed alcoves scooped out of the rock, providing an indication of the wind direction.

The wind may also carry a chemical along with it, usually from pollution, that reacts with the rock and alters it. This type of weathering caused the red color on the Redwall Limestone mentioned earlier. Sometimes, mere sunlight activates the process of alteration.

Sedimentary rocks are normally held together by a cementing agent, like the way cement holds together the grains and gravel of a concrete sidewalk, making it hard to deform or erode. But if these cement bonds are broken by an earth movement such as an earthquake, the loose particles are vulnerable and easily eroded.

A combination of erosion types banded together to produce the badlands topography of South Dakota and elsewhere. Here weak, poorly cemented sediments exposed to multiple freeze-thaw cycles, harsh sunlight, unrelenting wind, and normally dry conditions have resulted in "badlands topography." Recognized by its *dendritic* drainage pattern (*dendro* means tree, so this pattern resembles a tree with large branches, limbs and tiny twigs), mature dendritic drainage is thought to take many years to form with only meager intermittent rainfall. However, several unknown conditions must be assumed. Was the rock as hard then as it is today? Have rainfall and wind conditions been constant?

At Mount St. Helens, surprising things happened in those relatively brief, extreme conditions, demonstrating that a mature dendritic drainage pattern can develop almost overnight. The rocks there were softer, the temperatures were high, and the erosion forces were extreme, rapidly producing a topography just like badlands. And it remains there today, defying the present action of wind, rain, and freezing. Evidently, the standard geological theories of minimal change over long ages need to be revised.

Badlands erosion

Tectonic Forces That Shape the Earth

While erosional forces today usually apply locally to rocks and their environments, the effects of erosion can have an enormous extent. When much larger systems are considered, this falls into the catego-

Geologic Processes

ry of tectonic events. *Tectonics* involves crustal movements that affect large sections of the earth. Tectonic forces, likewise, have great power with which to change the earth and normally refer to processes such as earthquakes, mountain building, and continental movements.

The aftereffects of faulting within the crust can be readily seen as broken rocks, displaced strata, and straight-line cracks or gouges. A fault displaces the adjacent material and moves one side differentially with respect to the other. If there is no movement, the break is considered a fracture, merely a crack in the rock.

Faults are caused when stresses build to the rock's breaking point, exceeding the strength of the rock. Often the stress is accommodated by strain, or a bending or stretching of the rock, but there is a limit to how far the rock can flex.

Just as cold peanut brittle is hard, unable to bend or stretch, rocks that have become brittle will break when stressed. Under confining pressures, it is theorized that a rock can flex a great deal, but good experimental support for this is lacking and the idea runs counter to common sense. Rocks under pressure do undergo some deformation, but they cannot bend too much without breaking. On the other hand, soft, pliable rocks act more like saltwater taffy and deform much more without breaking.

Folds in rocks can frequently be seen, even in hard, brittle rocks. Tight folding in brittle rock shows that the rock folded while still in a soft, elastic condition. Otherwise, it would have shattered. When these folds are found in multiple layers at the same location, this indicates that all those layers were soft at the same time. This contradicts the standard assumption that strata are formed over long periods of time.

Plate movements are frequently thought to trigger tsunamis. Earthquakes or volcanic events affecting the ocean floor may result in a tsunami, also called a tidal wave. These events have nothing to do with tides, however, but form as a result of the tectonic forces imparted to the surrounding water by the sudden movement of the seabed. Water is virtually incompressible. As the forces are applied, they are absorbed by the water and transmitted through it. The resulting energy wave—not a water wave—often travels at great speeds that are undetected by boats floating on the open seas. When the energy wave nears land, where water depths are shallow, the wave rises, slams into the shore, and races inland to do great damage.

Similar energy can be imparted to the ocean by a meteorite impact, an earthquake with an epicenter near the ocean, or a coastal landslide—anything that involves a sudden jolt to the water.

In plate tectonics, as the continents collide, or as continental and oceanic plates collide, mountain chains might protrude upward due to compressive forces. Most continents contain sedimentary strata, which can crumple under collision. The Himalayan Mountains provide a good example of this type of boundary. A collision between a continent and an oceanic plate would send the denser ocean basalt subducting below the other plate, followed by volcanism along the margin, as is clearly seen in the Asian coastal regions.

Depositional Processes That Shape the Earth

We've already discussed the basic types of rock. Each involves different processes for their formation. The products of such processes are typically blanket-like layers, which then can harden into rock. Most of these processes involve moving water or wind, although volcanic eruptions also lay down strata.

The Deposition of Igneous Rock

We normally think of molten lava flowing down the sides of an erupting volcano. This hot, liquid material seeks a level, lower place to come to rest, much as liquid water does. Often the eruption occurs in pulses, so that the lava comes out in successive layers. These may harden at slightly different times and take on the look of several blankets stacked on top of one another. If the hot lava flows into water, which cools it quickly, it is called "pillow lava" after its pillow-like appearance.

The liquid lava contains a variable chemistry. As it cools, each mineral can form as separate crystals. The quicker the cooling or "freezing," the smaller the crystals appear, since they did not have time to grow. These rocks are known as *extrusive*. Rocks cool most slowly underground, since the overlying strata act as an insulating blanket to keep the heat inside. These rocks are termed *intrusive*.

Often some of these crystals contain radioactive atoms, such as uranium. These atoms decay over time, giving off radioactivity. For example, the large atom of uranium decays into the smaller atom of lead, releasing tiny subatomic particles. By measuring the precise amount of radioactive atoms, and by knowing the rate at which the element decays, it is thought that the age of the rock can be determined. Several unknown assumptions go into this calculation, however, which call the entire method into question.

To begin with, one must know (or assume) how much lead was present at the start. If any lead is present then, the rock already looks old even though it is really fresh. Next, it must be assumed that nothing has happened to add any uranium (or lead) to the rock or to have removed some of either element. If it has, the calculation is invalid. Finally, one must assume that the rate of decay has remained constant throughout the entire time since the rock froze. All of these assumptions are questionable and are known to often be violated. Radioactive dating (more properly called radioisotope dating) cannot be trusted and should not be considered authoritative.

Assumptions Used in Radioisotope Dating

- No daughter element (such as lead) was present at the beginning
- Nothing has happened to add more of the parent (uranium) or daughter (lead) element
- Nothing has happened to remove any of the parent (uranium) or daughter (lead) element
- The rate of decay has been constant since the rock was formed

Cardenas Basalt in Grand Canyon

Volcanic eruptions produce more than lava. Depending on the amount of volatile material mixed with the magma, ash may explosively erupt, be carried aloft, and fall to earth in "blankets." If caught in the wind, it can be transported for some distance.

Lava dome on Mount St. Helens

During the Mount St. Helens eruption, the ash cloud shrouded the landscape below in total darkness until the atmosphere cleared. Several states were somewhat darkened, and the weather around the world was affected for several years. The ash directly above the volcano remained aloft only as long as the jetting from below could support its weight. Then it crashed to earth and spread out as a devastating pyroclastic flow at several hundred miles per hour. The combined thickness of the rapidly deposited layers at Mount St. Helens reached 600 feet!

Many igneous rocks that cool at the earth's surface are categorized as basalt. These include the ocean basalts, which cover the ocean floor, and associated volcanic products. They are usually dark in color, richer in iron and magnesium, and quite hard. Resistant to erosion, they often make up vertical cliffs that are seen in many locations.

Another common igneous rock is granite, which slowly cools underground and is intrusive instead of extrusive. In fact, the bulk of the continental cores are thought to be composed of granite. These rocks are light in color and contain abundant quartz and feldspar minerals. The crystals in this granite had time to grow during the cooling process and can easily be seen. Granites are often found in large bodies and are usually not considered to be in the same category as blanket-like strata. Since they formed predominately underground, we cannot always observe layering, but in areas exposed by erosion we sometimes do see distinct layers. We know that molten rock can be squeezed into cracks and between layers, later solidifying into granite, but the history of many granite bodies remains a mystery.

Half Dome in Yosemite National Park is made of granite, and so are Mount Rushmore in South Dakota and Stone Mountain in Georgia. These granite bodies seem to have risen up as "blobs" from far below and occur with a series of similar blobs nearby. The precise conditions that led to these phenomena are not fully known. Many geologists believe that some granites formed near plate boundaries where the ocean crust was forced back into the mantle, creating granite-rich

Half Dome, Yosemite

Mount Rushmore

melts. However, the formation of huge, granite-rich continental masses remains a mystery, even in light of plate tectonics theory. Strange granitic formations, such as Devil's Tower in Wyoming, show vertical columns that formed when they shrank as they cooled and fractured in regular, columnar-looking patterns.

Sedimentary rocks that are composed of eroded and redeposited granite cover many areas of the continents. Sandstone grains are made of quartz, which comes from eroded quartz grains in granite. The same could also be said of other rocks. Shale is a fine-grained clay derivative of the silicate minerals (feldspars) in igneous rocks. Metamorphic rock, as we have seen, consists of these same rocks altered by heat and pressure. Scientists find it convenient to lump shale and other fine-grained rocks together into a new designation called "mud rocks." Typically thinly layered and laminated, they comprise the vast majority of the crust's sedimentary rock and are now known to be deposited under high-energy conditions of water flow.

The total density of the continental crust is similar to granite's density, but less dense than oceanic basalts that make up the oceanic plates. Being less dense, the plates containing continental crust, in effect, "float" on the upper mantle and rise to a higher elevation.

Coal seam, Utah. The vegetation that became this coal could not have grown in a swamp. Note the sharp contacts above and below the coal with the adjacent marine sediments.

Touchet Beds, Washington. Formed rapidly as an Ice Age lake burst through its natural barriers, eroding and then depositing much sediment.

GEOLOGIC PROCESSES

The more dense oceanic plates "sink" deeper into the upper mantle and are usually below sea level, while continents are usually above sea level. On a world scale of thousands of horizontal miles, vertical differences in elevation consist of just a few miles (from the ocean depths to the highest mountain summit).

The Deposition of Sedimentary Rock

Nearly all sediments require water (or sometimes wind) in order to be transported and deposited. Obviously, clastic sediments that consist of physical grains require more energy to move than do dissolved chemicals. A slow-moving stream might carry along smaller grains of clay, but the sand remains on the stream bed, seldom moving at all unless there is a greater than normal flow. Experiments have calibrated the water speed needed to move various grain sizes, including not only sand grains, but also cobbles and boulders.

Graded bedding at Split Mountain, indicating repeated pulses of sediment-laden water

Most sandstone layers cover great areas, sometimes semi-continental in scope, so the depositional process must also have been of great extent. Contrast this with the possible mechanisms for transportation with which we are familiar. Modern rivers move sediments and are known to meander, changing their location as they migrate across the river valley. Over time, they impact a larger area than their present stream bed. Is it reasonable, however, to claim they have migrated extensively in the past to cover great areas, with minimal vertical erosion, and are responsible for the extensive, horizontal sand beds known to geology?

A similar question could be asked about deltaic deposits. Modern deltas sport distributaries, with numerous higher and lower undulating spots—not at all flat-lying like typical sandstone beds. Sandstone beds could hardly be the result of past rivers or deltas.

Guyserite in Turkey, left behind when mineral-laden water evaporated

Much the same could be said about sandy beaches. Their rather narrow width does not match the enormous sandstone beds. Beaches can extend inland during storms and are theorized to have migrated as sea level changed, but this hardly explains the thick and extensive sandstones in the geologic record. Lake sediments accumulate as streams bring clastic material into the lake. Again, the scale is the problem. The streams transport very little material, and certainly not the amount needed to build up thick layers that span continents. Instead, the deep, wide lake bottom collects very little particulate material. Sand drops out of the current where it slows down near the shore, and only limited amounts of fine clay make it out to deeper locations.

Individual sand grains can also be moved about by the power of wind. Outcrops of granite in southern California are subjected to a strong prevailing wind from the west. They weather and disintegrate as the clay-rich feldspars weaken. The strong, resistant quartz grains, now loose, blow away and collect when the wind slows. Predictably, a large sandy desert, complete with migrating sand dune fields, sits just downwind of these outcrops. Once an area is covered with shifting sands, vegetation has a difficult time establishing permanent roots.

Dissolved chemicals that have precipitated from water to become sediments have many sources, some organic and some inorganic. Most often, limestones contain sediments from both sources. Limey sediments that have hardened into limestone may be derived from shells of sea creatures or reef fragments, or merely the secretions of sea creatures, along with material from inorganic sources, but all consist of calcium carbonate, $CaCO_3$.

GEOLOGIC PROCESSES

Limey sediments are being deposited in many places today, such as in the Caribbean Sea. They are always very fine-grained, composed of aragonite, while older sediments that have hardened into rock have recrystallized into the mineral calcite. Scientists wonder if the transportation and depositional processes in the past were much different than those today, since we find extensive layers of limestone in the geologic record.

Coal is an organic rock, consisting almost entirely of plant material. Geologists have long been taught that coal is altered peat that collects in the stagnant water of a swamp. More recent thought proposes lush vegetation growing near shore, but also possibly floating on shallow, coastal waters. As these organic-rich areas were uprooted and transported, their decayed remnants collected into vast layers underneath the former floating forest. No such forest grows today. The nature of coal formation remains a mystery, because modern peat swamps are so very different from the thick and extensive coal seams.

Depositional processes today also include the work of glaciers, and their geologic products are recognizable by distinctive landforms composed of "drift." Continental glaciers can still be found near both poles today, covering Greenland and Antarctica. Ice forms as falling snow collects and is packed into a sheet of ice. Gravity can move this sheet down even a meager slope. As it moves, it accomplishes much erosional work and transports rock fragments of all sizes trapped within the ice. Snow and, ultimately, glacial ice come from evaporated seawater. If all the glacial ice present in the world today were to melt, the sea level would rise several hundred feet.

Alpine or mountain glaciers similarly come from packed snow. They also erode, transport, and deposit sediments. The mountain glaciers in existence today are only a faint reminder of their number and size in the past.

Sediments do not necessarily turn into sedimentary rock, regardless of how much time is involved. Many geologic layers of sediments are still soft and unconsolidated, especially offshore. All sediments were soft when deposited, including those found

Mountain glacier

in modern sedimentary rock strata. Once in place, the weight of the overlying sediments pressed down, removing water and squeezing the individual grains together. Saturated sediments may take up 50 percent more volume than when the water is driven out. The final hardening process is really a process of chemical cementation. In either clastic or precipitated sediments, chemically precipitated mineral matter in the remaining pore spaces of sediments binds it together into a coherent mass. This can happen rapidly under the proper conditions. Just as concrete sets up in a matter of hours as the cement within chemically hardens, so rock lithification is dependent on a cementing agent. Often this agent, the "glue" binding it together, is calcite (perhaps from dissolved limestone) or silica (perhaps from dissolved sand) moving in solution in the groundwater.

Higher temperatures speed up almost every process, and this is true for most of the steps in rock hardening. Hot water flowing through limestone will accelerate dissolution of calcite, and if flowing through sandstone will carry off dissolved silica. As the water percolates through porous sediments and the chemicals precipitate, cements are formed, spanning the space between the individual grains. As the water is driven off, or chemically incorporated within the cement, the grains transform into solid rock.

A similar cause is believed to produce the process of fossilization. Each type of fossil requires rapid burial, the absence or limiting of scavengers, and bacterial action. Few fossils are forming today, and these only due to rapid burial in local catastrophic events.

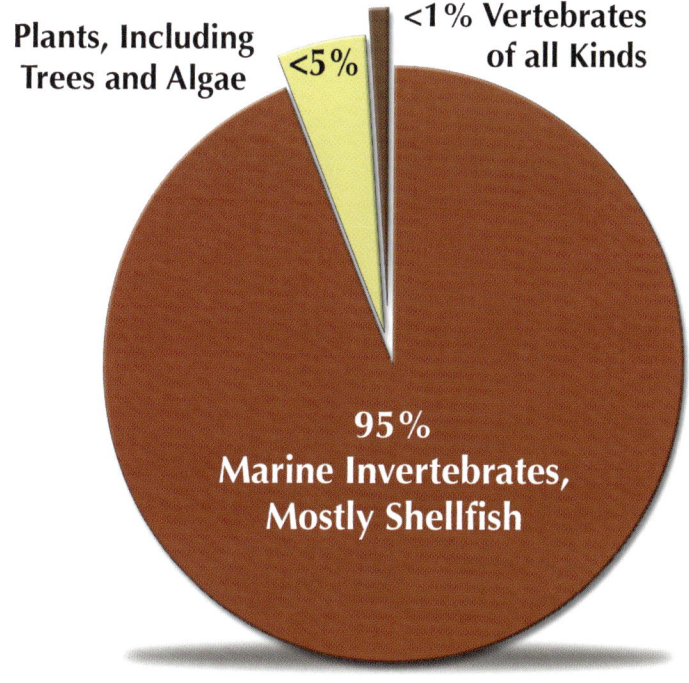

Abundance of fossils

Fossils are preserved in a variety of ways and forms.

- Replacement of organic material
- Cast of original shape
- Hard parts preservation
- Soft parts preservation
- Impregnation or petrification
- Carbonization
- Tracks or burrows
- Frozen life
- Coprolites or gastroliths

In order to be fossilized, a plant or animal must be quickly buried, but the rapid processes do not stop there. Contrary to common belief, it does not take a long period of time for something to fossilize—it just takes the right conditions. Chemicals in percolating groundwater, elevated temperature, and injection pressure—all contribute to rapid fossilization. There are numerous examples of remains that have been quickly altered to become a fossil, in even a few months.

Oceanic Processes That Shape the Earth

Perhaps the most powerful processes on the earth's surface are oceanic processes. Certainly the oceans are huge, occupying about three-fourths of the planet's surface. Much of the ocean depths are as yet unexplored. Life of all kinds abounds within it, and more species are discovered each year. The ocean waters themselves contain various dissolved chemicals that react with its contents and with the land beneath.

When we think of the ocean, we often picture billowing waves pounding the shore. Their unrelenting action can accomplish great geologic work. Waves disperse animals and plants of a wide variety, keeping the food chain intact. Ocean currents have been mapped and constitute virtual highways in the seas. They not only facilitate shipping efforts, but create good migration lanes for fish and other marine creatures, and serve to moderate climate. For example, Great Britain sits rather far north on the globe, but enjoys a moderate climate because of the warm, moist Gulf Stream current that flows northward from the Gulf of Mexico.

Tides are different than waves. Tides result from the pull of the moon and sun's gravity, but mostly the moon's. Although about one-quarter the size of earth, the moon acts almost like a twin planet, and its gravitational pull is significant. As it revolves around earth roughly once a day, water is attracted to it, creating a bulge that follows the moon's path. High tides usually occur twice a day, once as the moon is directly overhead and once as it tugs from the opposite side of the earth.
Tides exert a major influence on the shoreline and inlets, repeatedly flooding and draining susceptible areas.

Waves, on the other hand, are caused almost solely by wind. The stronger and more sustained the wind, the greater the wave height. Wind and waves crash into coastlines, weakening and eroding the land. Storms accelerate and accentuate the process. Hurricanes are generated far out to sea by temperature differences, but can travel far inland and level natural formations and human structures alike.

In recent years, meteorologists have proposed the possibility that extremely large hurricanes, known as hypercanes, occurred in the past.

Satellite view of Hurricane Katrina

GEOLOGIC PROCESSES

Their formation requires warmer oceanic temperatures and colder continents than exist today, but seem to have been necessary to account for certain ancient features discovered on land. What must the past history of earth have been like?

Hurricanes are not the primary model for past deposition of sediments. Rather, they are more adept at erosion. Much the same could be said for tsunamis. The "Christmas" tsunami of 2004 eroded beaches and uplands, but left only minor sedimentary deposits.

Another physical process that bears inclusion is meteorite impact. On a daily basis, earth encounters numerous sand-sized particles from outer space, but these burn up in the atmosphere, with only fine dust reaching the surface. Ice particles also enter from space and their vapor adds to the total water on the planet.

Creation and the Genesis Flood

Many of the geological mysteries mentioned so far can be explained when we take into account the biblical record of the earth's early history. Scripture tells of past events that really happened at God's hand. It first speaks of creation, a unique period before today's natural laws had been instituted.

In the beginning God created the heaven and the earth. (Genesis 1:1)

The first act of creation was the creation of the heavens and the earth. Note that this involved space (the heaven), mass (the earth), and time (in the beginning). Physicists will recognize the reference to the space-mass-time universe, which encompasses all of reality. The triune universe is a reflection of the triune Creator. Of particular interest to this study is the creation of physical matter, out of which the physical universe was molded. Never again does the Bible make note of the creation of mass. The oceans and atmosphere were formed on Day Two. The continents and plants were made on Day Three; the sun, moon, and stars on Day Four; sea life and flying creatures on Day Five; and land animals and humans on Day Six. All were fashioned of materials that God had created (at least in principle) on Day One.

The second act of creation occurred on Day Five. For the second time, something entirely new came into being as God created living creatures. The physical bodies of the animals were of the same material He created on Day One, but He added consciousness to the inanimate matter—thus, living creatures appeared. This consciousness might also be called "the breath of life."

The third and final unique act of creation was on Day Six, when God created humans in His own image. Humans possessed the same life principle as the fish, birds, and land animals, but in addition they were given a new thing, the image of God. This was man's spiritual side, his conscience, love of beauty and justice; i.e., his eternal spirit. Man is not God, but something in mankind, together in the combined male and female, sufficiently reflects God's nature so that He can call it God's image. We do not have God's power, omnipresence, or omniscience, but in our original created state we reflected Him adequately.

This triad of physical nature, conscious life, and a spiritual side can also be understood as body, soul, and spirit. The terms are used somewhat interchangeably in English and in Scripture, but couching God's image in man in this way is instructive.

Considering the earth's makeup, the events on Day Three are of enormous importance. On this day, the oceans were gathered into "seas" and the dry land appeared. Gravity in some form was already operating, and as the waters rushed off the emerging continents, it likely would have eroded the land and deposited material in the ocean depths. Erosion today is destructive, but under controlled conditions can do useful work. One might assume it produced underground aquifers and drainage channels to be used permanently by water sources and rivers, and that the eroded sediment deposited offshore rapidly formed beaches filled with nutrients for the life that was to be created later that day. Remember, there was no life yet created, so all sediments would have been completely fossil-free.

Day Four saw the making of the astral bodies. The sun today is the source of the energy used by plants and animals. The moon allows the tides to continually refresh the seas and facilitate life.

Days Five and Six witnessed the creation of life, first in the seas, then in the skies, and then on land. The rooting of plants, the burrowing of earthworms, and reworking by animals further alter the land, doing continual geologic work. All in all, creation was "very good," with all things in balance, continually recycling energy and products.

The Great Flood of Noah's Day

Soon, however, man's disobedience ruined God's very good earth, over which man had been placed in charge. Adam and Eve's sin brought death and decay on all of their dominion. And now everything dies. Man dies and so do all animals. Not only that, the sun is burning out. The moon's orbit decays each year. Indeed, all of creation has suffered from man's tragic choice. Only man can sin, but man's sin brought chaos to God's masterpiece of creation. Before long, man's wickedness was so great that God's patience was at an end. He promised the wages of sin and sent the Flood as His instrument of judgment. The Flood's primary purpose was the punishment of mankind's sin. But it also fully altered all of man's home. It totally restructured the earth's crust.

Employing waters that God had stored in the "windows of heaven" (the atmosphere) and the "fountains of the great deep" (underground aquifers), the Flood destroyed the entire earth's surface. Scripture implies mega-volcanoes and mighty earthquakes. These would necessarily have resulted in massive tsunamis. The Flood itself could probably be termed a complex of hypercanes, pouring unimaginably huge amounts of rain on the land below.

40   Geologic Processes

Think of the erosion that would result from such a worldwide torrent. This would create sediment deposits in low places, the ocean basins, and valleys. These layers would also be filled with formerly living things, awaiting fossilization.

The continents, though unable to be moved by normal geologic forces, were free to move given the much larger forces employed by the Flood. The dense oceanic plates dove beneath continents, while mountains were pushed up by continental collisions. Seismic surveys have "found" pre-Flood ocean basins near the core mantle boundary, having rapidly dived and dragged the continental plates along behind them. Creation scientists acknowledge the impossibility of continental "drift"; they talk of continental "sprint" instead, as the continents broke apart and were driven toward their current locations.

The catastrophes that marked the Flood were followed by residual catastrophism as the earth struggled to regain its equilibrium. The movement of the continents would have ground to a near halt in the positions that they occupy today. Over the next few centuries, earthquakes would have tapered off and volcanic eruptions would have become progressively less intense.

Following the Flood, the oceans would have been abnormally hot due to the rapid formation of new oceanic crust, the superheated waters of the great deep coming up from below, and frictional heat from tectonic movements. The atmosphere would have been filled with volcanic debris, blocking incoming sunlight.

In those early centuries following the Flood, earth's systems were out of balance. Now we see them operating with regularity and in near equilibrium. The jet streams are predictable, the ocean currents follow regular paths, and seasonal changes come one after the other. It probably took several hundred years after the Flood to reach this stability. The Ice Age hit with a vengeance during this time of readjustment. It was caused by excessive evaporation due to elevated ocean temperature. The newly exposed continents were initially barren of vegetation and colder than now. The difference in temperature between continents and the oceans would have been great, transporting moisture to the land, where snow would have rapidly accumulated. Due to a cloudy and dusty atmosphere, the snow

GEOLOGIC PROCESSES

41

would not melt in the summer. As it built up, the snow would pack into hard ice sheets, covering extensive parts of North America, Europe, and Asia. This situation would continue until the oceans released their excess heat and the atmosphere cleared as volcanism waned.

In general, we can say that the world is actually a museum of past processes that operated on a much greater scale, proceeded at a more rapid rate, and acted with more intensity than the processes in force today.

Conclusion

These were the things I included in my talk before the Chinese leaders. At Mount St. Helens, local catastrophic conditions accomplished much rapid geologic work of the same quality, although on a smaller scale, and of the same nature that students are taught require long periods of time. I pointed out that I had not learned these truths during my education. They were censored from the students. Because of censorship, I had been poorly educated, incapable of competent geologic scholarship.

Communist Party officials knew that I spoke of them and considered it a major insult, and called for my arrest. Thankfully, cooler heads prevailed. Surprisingly, when the meeting closed, the government's chief scientist, the cabinet head of their Ministry of Science and Technology, confessed ignorance of the dynamic processes I had mentioned and asked if I could visit some geologic sites with him and give lectures to the science community. He could see the potential benefit of learning this censored information.

The record written in the stones around us provides abundant evidence that the earth history recorded in the Bible is true and can absolutely be trusted. As can the Creator of that earth, whose Word the Bible is and who alone offers redemption to a world lost in sin. For "neither is there salvation in any other: for there is none other name under heaven given among men, whereby we must be saved" (Acts 4:12). And one day He will return for His own and put in place a new heaven and a new earth, where His people can once again live in perfect fellowship with Him.

Bibliography

Austin, S. A. 1995. *Grand Canyon, Monument to Catastrophe.* Santee, CA: Institute for Creation Research.

Coffin, H. G., R. H. Brown and L. J. Gibson. 2005. *Origin by Design.* Hagerstown, MD: Review and Herald Publishing Association.

Morris, J. D. 2000. *The Geology Book.* Green Forest, AR: Master Books.

Morris, J. D. and S. A. Austin. 2003. *Footprints in the Ash.* Green Forest, AR: Master Books.

Morris, J. D. 2007. *The Young Earth.* Green Forest, AR: Master Books.

Morris, J. D. and F. J. Sherwin. 2010. *The Fossil Record.* Dallas, TX: Institute for Creation Research.

Snelling, A. A. 2009. *Earth's Catastrophic Past.* Dallas, TX: Institute for Creation Research.

Image Credits

Dr. John Morris: 16, 17, 23, 31 (right and lower), 32, 33

NASA: 13, 37

U.S. National Park Service: 21

U.S. Geological Survey: 10, 11, 18

Susan Windsor: 15, 35

Science Education Essentials

How do I explain the differences between biblical creation and evolution?
What evidence for the origin of life should my students know?
Where do I go for trustworthy information on science research and education?

For 40 years, the Institute for Creation Research has equipped teachers with evidence of the accuracy and authority of Scripture. In keeping with this mission, ICR presents Science Education Essentials, a series of science teaching supplements that exemplifies what ICR does best—providing solid answers for the tough questions teachers face about science and origins.

This series promotes a biblical worldview by presenting conceptual knowledge and comprehension of the science that supports creation. The supplements help teachers approach the content and Bible with ease and with the authority needed to help their students build a defense for Genesis 1-11.

Each teaching supplement includes a content book and a CD-ROM packed with K-12 reproducible classroom activities and PowerPoint presentations. Science Education Essentials are designed to work within your school's existing science curriculum, with an uncompromising foundation of creation-based science instruction.

Demand the Evidence. Get it @ ICR.

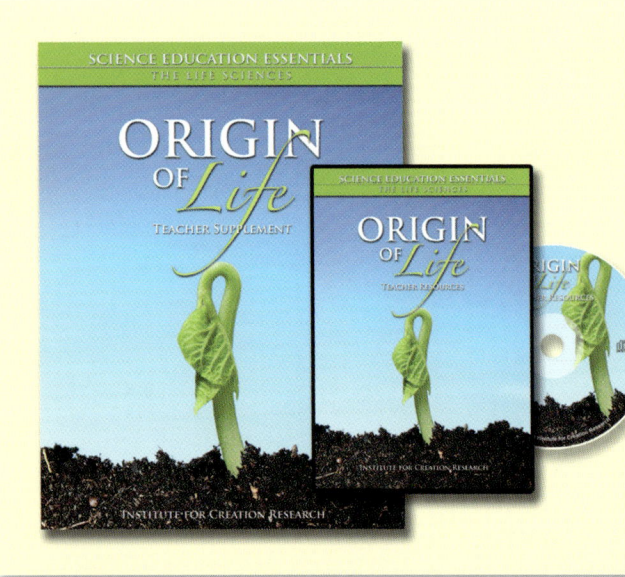

Origin of Life

How did life get started on earth? Many scientists believe that life began from natural processes, but the Bible presents an alternate explanation.

Origin of Life, the first of the series, answers basic life questions such as:

- What is the origin of life?
- What are the physical and biblical definitions of life?
- What are the physical requirements for life?
- Can life exist elsewhere in the solar system?
- And much more

It gives scientific explanations for the chemical basis for life from a biblical worldview and discusses the efforts to create life in the laboratory. Most importantly, it offers scientific evidence proving that the creation of life "requires an act of God."

Available for **$24.95** (plus shipping and handling)

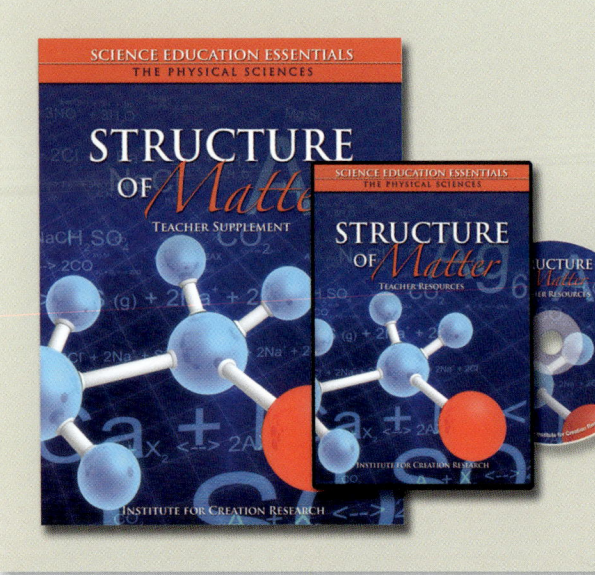

Structure of Matter

Predictions in science are based on knowledge of observable events. The accuracy with which science can make predictions points to the order and structure God established within His created universe.

Structure of Matter, the second of the series, explores structural forces and elements of nature such as:

- The First and Second Laws of Thermodynamics
- The structure of the atom
- The periodic table
- Properties of matter
- And much more

The order and design of the universe point to a Creator of omnipotent power and omniscient strength. Truly, the structure of matter upholds the truth that "in the beginning God created the heaven and the earth."

Available for **$24.95** (plus shipping and handling)

Human Heredity

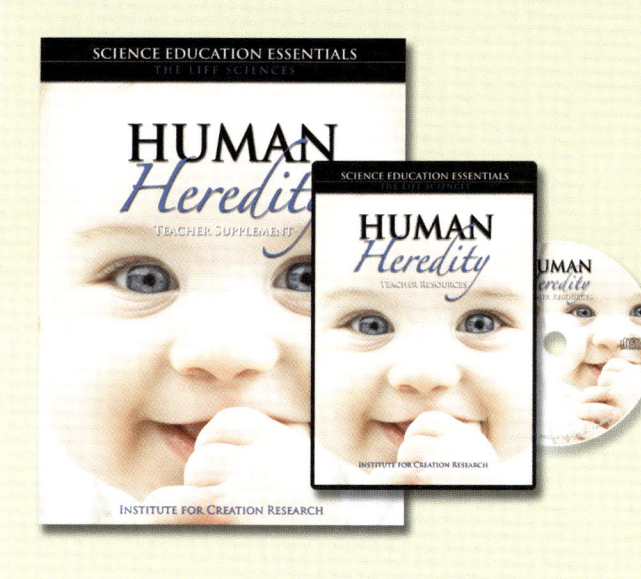

Genes provide most of the information that determines physical appearance and even influences certain behaviors. In spite of the differences among humans, their genomes are still 99.9% identical. Did everyone come from two people?

Human Heredity, the third of the series, examines such topics as:

- Our inheritance from our parents
- Dominant and recessive traits
- Human descent from Adam and Eve
- Polygenic inheritance
- And much more

The study of genetics has expanded our understanding of human inheritance, leading to the inevitable conclusion that all humans came from the first created man and woman. "And Adam called his wife's name Eve; because she was the mother of all living" (Genesis 3:20).

Available for **$24.95** (plus shipping and handling)

Genetic Diversity

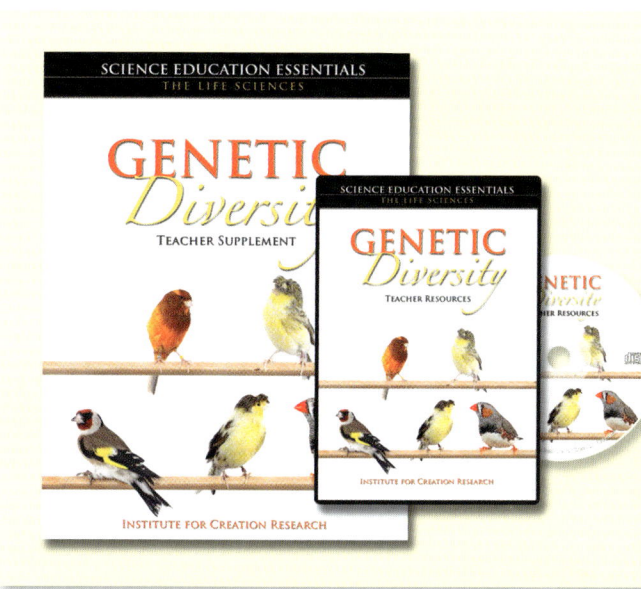

God created an incredible variety of incredible creatures—and it seems He created us in His image to enjoy that variety. What is the science behind this wonderful diversity?

Genetic Diversity, the fourth in the series, takes an in-depth look at:

- The classification of living things
- Differences among species and within kinds
- Diversity and the mosaic concept
- And much more

When it comes to the kinds of life observable today, the overwhelming scientific evidence strongly supports the thinking of creation scientists that life can only come from the eternal Creator.

Available for **$24.95** (plus shipping and handling)

Geologic Processes

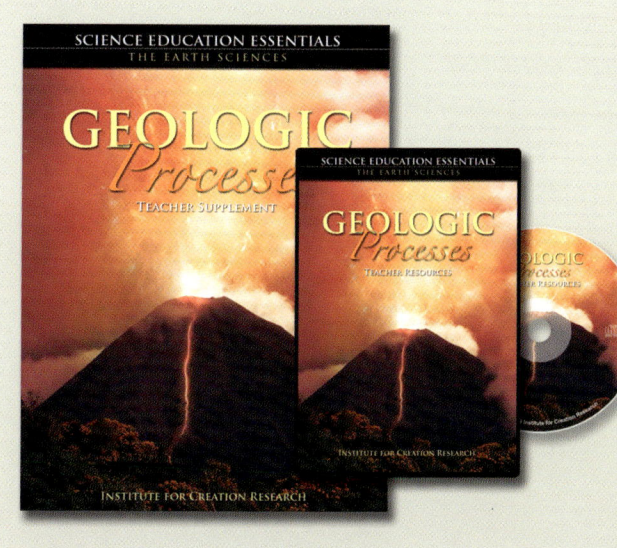

What geologic processes shaped our earth? Is evolution right, that it developed gradually over millions of years? Or does the geologic record demonstrate something else?

Geologic Processes, the fifth in the series, studies earth's history to answer the questions:

- Did the earth start as a cosmic collection of star dust?
- What processes shape the earth today?
- What types of rocks are found on earth?
- What is the geologic evidence for a worldwide flood?
- And much more

The world is a museum of past processes that operated on a much greater scale, proceeded at a more rapid rate, and acted with more intensity than those acting today. The best explanation for earth's history is the biblical record of creation and the great Flood.

Available for **$24.95** (plus shipping and handling)

For More Information

Sign up for ICR's FREE publications!

Our monthly *Acts & Facts* magazine offers fascinating articles and current information on creation, evolution, and more. Our quarterly *Days of Praise* booklet provides daily devotionals—real biblical "meat"—to strengthen and encourage the Christian witness.

To subscribe, call 800.337.0375 or mail your address information to the address below. Or sign up online at www.icr.org.

Visit ICR online

ICR.org offers a wealth of resources and information on scientific creationism and biblical worldview issues.

- ✓ Read our daily news postings on today's hottest science topics
- ✓ Explore the Evidence for Creation
- ✓ Investigate our graduate and professional education programs
- ✓ Dive into our archive of 40 years of scientific articles
- ✓ Listen to ICR radio programs
- ✓ Order creation science materials online
- ✓ And more!

For more information, contact:

Institute for Creation Research

P. O. Box 59029
Dallas, TX 75229
800.337.0375